The Best Fences

James FitzGerald

CONTENTS

Introduction

Although this bulletin concerns itself principally with the practical aspects of fences, don't overlook the fact that you are working with one of the most profound developments in the history of mankind. Next to the wheel and the plow, the fence ranks high as a major landmark in the progress of civilization. Historically, it was such a simple tool and yet it gave man the enormous capability of making boundaries, separations, edges, and limits. For the first time, a man could organize his geography and the things in it. He could keep things in and keep things out. He could contain his livestock, defend himself with barriers, and define his territory.

Fences evolved from a simple enclosure for domesticated animals in 4000 B.C. to a massive statement of power in the form of the Great Wall of China which extended 3,000 miles in 300 B.C.

In addition to being a mere physical boundary, a fence carries a genuine and deep psychological significance. Standing as a crude wooden sculpture between fields or as an elaborate wrought iron entrance to a chateau, it symbolizes the basic way we handle the vastness of life. It fulfills our ancient human need to create manageable units out of chaos. Not only does it afford us a way to organize our physical spaces, but more fundamentally, it reflects the primal way we define our psychological spaces; mine / yours; me / you.

A fence simply lets us define ourselves and our world. It is a monument to Order.

Mistakes

The best way to avoid mistakes is to know where they are hiding. In the list below I have defined eight of the major areas in which big mistakes occur. Sections of this bulletin deal with each of these potential problem areas.

- How to plan ahead.
- How to construct a legal fence.
- How to avoid fence post problems.
- How to choose the right type of fence.
- How to save time and energy by using the right tools.
- How to use the best construction technique.
- How to construct workable gates.
- How to build safely.

How to Plan Ahead

Four basic indoor planning steps and an equal number of outdoor planning steps are needed to keep you away from mistakes or miscalculations in building a fence.

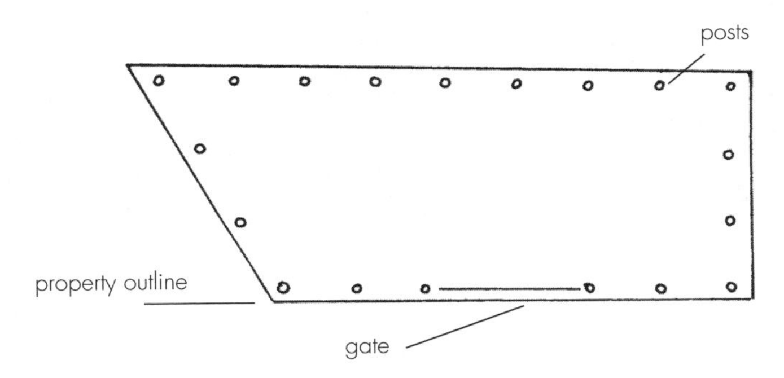

1. Draw a map of your property on ¼-inch graph paper (¼ inch equals one foot).

2. Using a ruler, sketch in the contemplated fence line. Try to keep the scale accurate.

3. From your anticipated traffic pattern, locate the gate area first. Mark out the gate posts, then the corner posts, and finally each line post. If you find that the distance between gate post and the corner post is not evenly divisible by eight feet or six feet or whatever your section length will be, you have two choices.

a. Shorten each section a bit so that all sections have identical lengths.

b. Plan for a shorter section next to each corner post. This second option is usually preferred because it wastes less time and materials. It would be very time-consuming if every eight-foot board, for instance, had to be trimmed by 7¼ inches. Also, these short sections make excellent braces for the corner posts.

4. Write down the various lengths. For the first time in your life you may be confronted with working in rods. Pasture land boundaries are frequently measured in rods, and rolls of barbed wire always come in rods. Don't panic. A rod is simply 16½ feet.

From the above data you can calculate the answers to these basic questions.

a. How many corner posts, bracing posts, and line posts will you need?

b. How many boards, pickets, rails, or bales of wire will you need?

c. How many gates and gate posts will you need?

Once satisfied with your graph paper version, go outside and recreate your ideas on the ground in the next four steps.

1. Place a stake at the site of each gate post. By starting your fence at each side of the main gate and proceeding with standard size sections toward the corners, you can maintain perfect symmetry around the gate area where fence asymmetry or other defects would be most noticed.

2. Place stakes at the corners.

3. Pull mason's twine tightly between the gate posts and the corner posts after clearing any brush or obstacles that may be in the way.

4. Lay a board or rail or any unit of horizontal "stringer" material along the twine to locate each successive line post location. Mark each spot with a stake.

In this planning phase, don't be shy about asking for help. People who own beautiful fences are very proud of them. They are usually delighted to give you all of the details and help that you want. So if you see a terrific fence, ask about it.

In the back pages of this bulletin, you will find a list of other good fence books. Most of them are full of beautiful photographs and details that may make your job much easier.

Most of the giant catalog stores have at least a small fence section mixed in with garden supply or hardware sections. The prices for materials are usually quite good, especially for items such as pickets, should you like their design. The amount of information you can obtain about fencing construction varies with the salesperson. If you want to talk fencing, the local hardware store personnel are usually unbeatable. They know the local soils and weather conditions, and have a lot of experience with do-it-yourself problems in your area.

If you need professional help with the planning or construction, look in the yellow pages under landscape design, contractors, or fencing materials. Even if you only want help with your plans, ask about estimated costs for their time and services.

How to Construct a Legal Fence

First, make sure you are putting the fence on your own property. It is embarrassing to construct a fence on someone else's land, and it can also result in a loss of ownership. Consult your own survey maps or verify your boundary lines at the recording clerk's office. Find out if there are any restrictions such as height limits, set-back rules, or construction codes.

If you plan to run the fence exactly on the boundary line between your property and your neighbors', talk with them about it. Even if you fear the possible response, try it. You may get more cooperation than you expect. There are many possible responses. Here are some suggestions for handling the two extreme positions.

1. Your neighbor may be willing to share construction and maintenance costs, thereby effectively flipping Robert Frost's statement into good neighbors make good fences. If you and your neighbor agree on certain aspects of fence construction, write down what you agree upon so that the details don't get distorted as time goes on.

2. If you are not fortunate enough to have a concurring neighbor, make sure your fence is located a foot or so inside your property boundaries so there is no ownership dispute.

In addition to laws that may restrict your fence variables, some laws may require you to build fences. Frequently, it is mandatory to construct a barrier to attractive nuisances such as swimming pools and excavation sites.

How to Avoid Fence Post Problems

Most fences share a need for posts every few feet to support the wood or wire in the horizontal sections. The classic Virginia zigzag rail fence is one of those exceptions; it does not require posts when constructed with angles less than 135 degrees. Nevertheless, the post is the most common denominator to all fences. It will become the backbone of your system, and it demands careful selection. A good fence post should demonstrate three basic characteristics: Stability, survivability, and straightness.

Stability

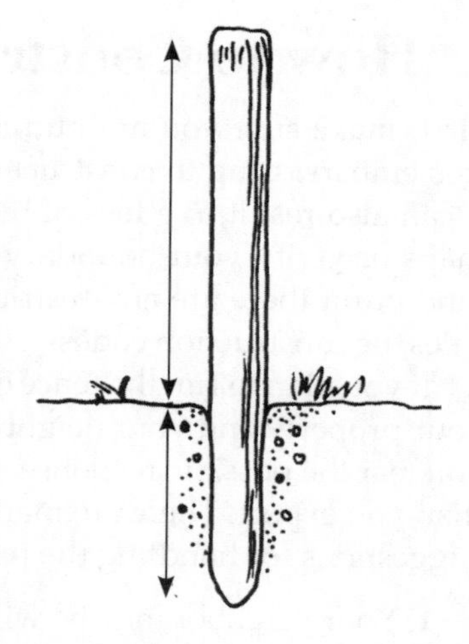

Generally, one-third of the fence post should be well anchored underground. If you want a four-foot post above ground, start with a six-foot post. Before you dig the hole, it pays to check out a couple of things. Avoid ledges and rocks, buried cables and water pipes, leach fields and septic systems.

A post often can be set without even digging a hole. Thrust a long iron bar downward several times in the same spot. It should go a little deeper each time. After it penetrates the ground a foot or two, grasp the top of the bar with both hands and rotate it as if you were stirring a witch's cauldron. Next place your nicely sharpened fence post into the earthen tunnel you just created and pound it in. You finish with posts that are held tightly by compressed soil, and avoid having to fill in a large hole.

Pounding the Posts

Be careful when you are pounding the post into the ground. If you are using a sixteen-pound sledgehammer or a maul, make sure the head is not loose and the handle is not cracked just below the head. Position yourself so that the head of your sledge hits the post at approximately the level of your waist. Unless you are eight feet tall, you should stand on something stable such as the back of a pickup truck to start each post. Face the post directly, spreading your feet apart (so if you miss the post the sledge will be headed between your feet and not at one of them), get balanced, and look exactly at the spot you want the maul to hit the post. Don't blink. If you blink just at the moment of impact or a fraction of a second before, you risk splitting the post, cracking the sledge handle on the top of the post, or smashing some toes. Watch carefully as you are striking the post, don't blink, and wear protective glasses.

Planting the Posts

If you can't punch a hole with an iron bar, you can plant your posts. Making the hole may be tedious but it is simple. Use an iron bar to loosen the dirt and rocks, and a shovel to empty the hole. The advantage of this method is that you do not need sharp posts and you can construct a post enclosure that is well drained and allows the post to be solidly wedged with rocks and dirt. Frost is less likely to heave this post.

Using Concrete

For an extremely solid post, a wide ring of concrete at the base may be built. Place the concrete in an open collar around, but not under, the post so that the moisture can drain down along the post as the post shrinks or decays with age. Avoid sticking the post in concrete as if you were sticking a candle in frosting. This creates a moisture chamber around the post and tends to hasten its deterioration.

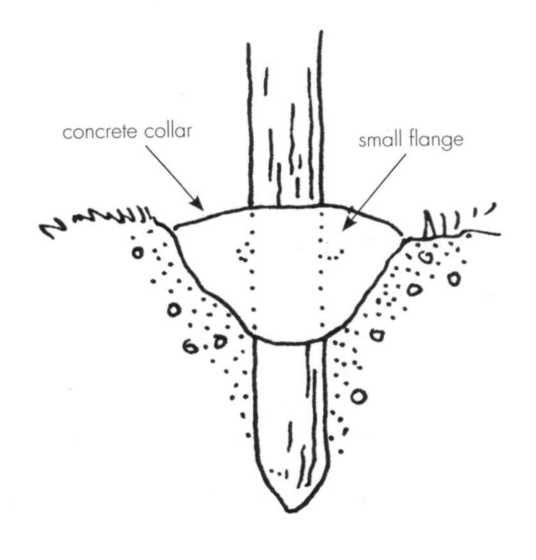

Mix the concrete a little bit on the dry side. Use two parts of cement, three parts sand, and five parts gravel with just enough water to make the mixture fluid but still somewhat firm.

If you use concrete, make the top of the hole two or three feet wide, mix the concrete, put the pole in position, and pour the concrete around it. To increase the bond between a post and the concrete, nail small flanges or cross pieces to the bottom of the post before putting it in position. Smooth the surface of the concrete so that the highest point is next to the post. This allows the water to drain away. It is also a good idea to let the concrete cure for a day or two before you use the post.

Durability of Untreated Heartwood

Resistance	Tree Type	Life Expectancy
Extreme Decay Resistance	Western Juniper Osage Orange	20–30 years
Good Decay Resistance	Sassafras White Oak Red Cedar Cherry	10 years
Poor Decay Resistance	Birch Beech Ash Elm Hemlock Hickory Maple Red Oak Poplar Willow	2 years

Survivability

It would be great to find a post that would resist rot or rust forever, but there aren't any. Post rot usually occurs at the ground level because that is where the combination of food supply, oxygen, and moisture is ideal for the growth of fungi. Woods decay at different rates, depending on the type of soil they stand in and on the amount of natural fungus-retarding chemicals present. The core of every cylindrical tree trunk (heartwood) contains more of these chemicals than the remaining wood (sapwood).

This table describes the hierarchy of preferable fence post heartwood.

The lifetime of fence posts can be lengthened. Peeling them is one method. Some oldtimers believe that posts set in the ground upside down rot sooner.

"Weak" wood such as maple can be converted into a "strong" wood by a variety of techniques. A simple method that offers limited protection consists of placing the lower half of the fence post in a fire long enough to produce a layer of char.

In the past, builders of fences have soaked the posts in either penachlorophenol or creosote. Now the U.S. Environmental Protection Agency has placed restrictions on the use of these, limiting it to persons who have passed a course on the safe handling of hazardous materials, and restricting sale of these to these certified applicators.

As a result, we must recommend that you use copper naphthenate, which long has been recommended for use around gardens, since it will not kill plants, as the other two will. Follow the directions on the container. As with all wood preservatives, the wood is best protected if it is dipped into the liquid, rather than being painted, if the work is done in the shade, and if the wood is clean and, particularly, unpainted.

Straightness

Make sure that the posts are standing straight and in a straight line. This is important cosmetically as well as practically, particularly when using wire fencing, because any post that is out of line is subject to enormous stress. If it bends over, the wire along the entire fence will be loosened.

To get each post standing straight, use a carpenter's level.

To get the posts in a line, run two pieces of twine between two corner posts. Tie one at the bottom of both posts, the other near the top. Place each post so that it just touches each string, keeping all posts on the same side of the strings.

Bracing

If you think that one post will encounter more horizontal stress than the others, brace it. All corner posts and gate posts should be braced, with corner posts braced in two directions. Here's how to do it.

1. The anchoring post should be larger than the other posts and should be set deeper, if possible. Try to bury about 3½ feet of an eight-foot post.

2. Set another large post four to six feet down the line.

3. Place a strong horizontal post between the two posts about a third of the way down from the top of the posts. This horizontal post should be set in a dado joint, or at least blocked for stability.

4. Circle eleven-gauge wire twice around these posts, running it from the top of the bracing post to the base of the corner post. Twist the ends of the wire together and staple it to the post.

5. Insert a strip of wood at the midpoint of the bracing wire and twist it like a tourniquet to tighten the wire.

6. Anchor it with wire against the horizontal bar.

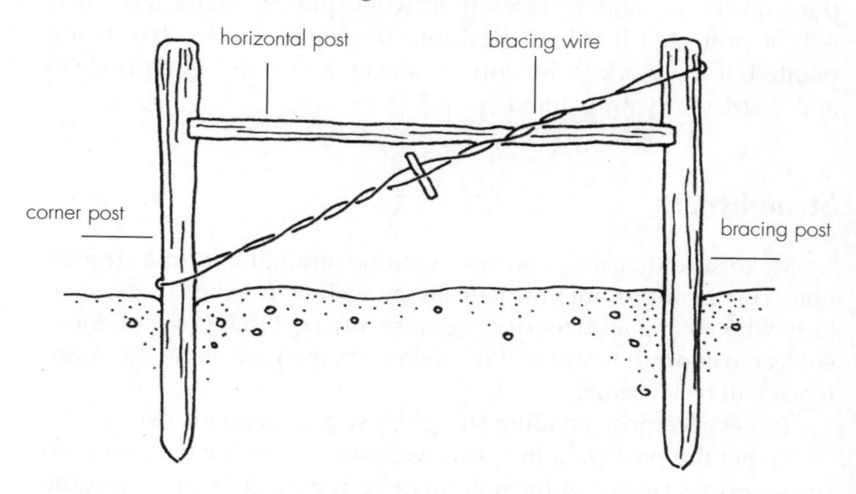

The physics of this bracing system is simple. Any force tending to pull the top of the corner post down the fence line is indirectly shared with the solid base of the same post. Through triangulation with a rigid horizontal bar and tight wire, two simple posts have been changed into a system of counterlevers. Remember, if you run the wire in the opposite direction (from the top of the corner post to the base of the bracing post), it will be useless.

How to Choose the Right Type of Fence

There are almost as many kinds of fencing materials as there are varieties of ice cream. The selection depends principally on which materials are the most readily available, least expensive, and will do the job properly. A five-strand barbed wire fence may be great for containing cattle but it just won't do around a school yard. The table gives you an idea of the popular options.

Virginia Rail Fence

Virginia rail fences or zigzag fences were common as the United States was changing from woodland to farmland. Trees cut during the clearing of fields and pasturelands were in abundance. When laid in the proper fashion, this type of fence could be constructed quite easily because it required no posts. It was also a good choice for pastureland because of its mass and sturdiness. Occasionally, a vertical post would be used at each angle to assure alignment and stability, but in general very few vertical posts were used. This type of fence can still be seen as an aging monument to classic fencing beauty reminding us of an age when materials were in abundance and inexpensive. Very few new zigzag fences are built because they consume an enormous amount of expensive material and also because their rambling design wastes acreage. Nevertheless, it is a distinctive fence and may be exactly the statement you want to make.

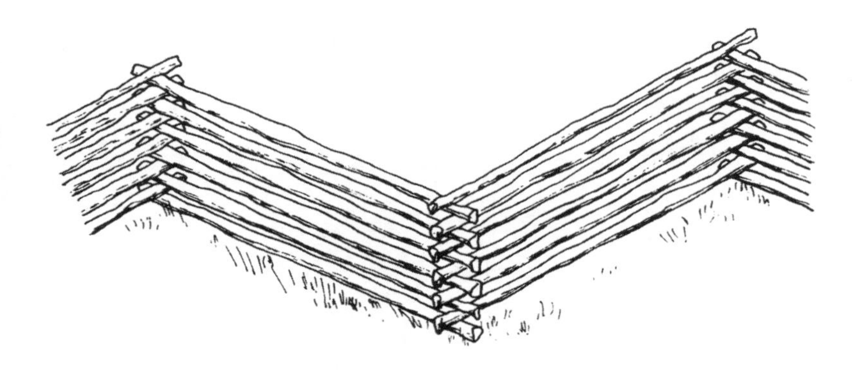

Fence Varieties	
Requirements	**Favorite Options**
Decoration	Virginia Zigzag Picket Post and Rail (two-rail variety) Boards
Deflection of Wind, Noise	Solid Boards
People	Chain Link
Domestic Animal Management:	
Cows	Wire: Barbed, Electric, or Woven
Horses	Boards
Sheep, hogs, goats	Woven Wire or Electric

Split-Rail Fence

A much more economical approach to the use of material came with the development of techniques for splitting timber into rails. Two- or three-rail fences are handsome and serve as very sturdy enclosures. Unfortunately, the posts need extra work. Holes must be chiseled or drilled through the fence posts at heights corresponding to the desired rail levels. To avoid the tough job of making holes in fence posts, an alternative method is to use two posts at each position instead of one. A horizontal connection between the two posts is made with a pin or a dowel to hold up the rails.

This type of fence will be more expensive than woven wire, barbed wire, or electric unless you can make your own rails.

The Board Fence

With the increasing availability of machined boards and nails, the board fence has become more popular. Painted or unpainted, a board fence is the aristocrat of the fencing world. It can be seen surrounding horse pastures or dressing up estates like an elegant frame on a masterpiece. Horses will need a six-foot high fence with fewer horizontal boards. A decorative fence need not be as high.

The cost of board fencing will be more than for split rail or electric fencing but probably less than woven wire fencing.

Barbed Wire Fence

The invention of barbed wire was one of the major events of the 19th century. Joseph Glidden gets a lot of the credit for the invention, but the first patent was given to Michael Kelly in 1868. Glidden popularized barbed wire by perfecting an easy way to manufacture it. By 1900 there were at least 1,000 different designs with many clever wire weaves and points. Old barbed wire "cuts" (eighteen-inch lengths) are extremely valuable collector's items today. Despite that history of a large variety of types of barbed wire, today there are fewer than six standard styles. A 16½-gauge two-point wire is the favorite for cattle owners, who use three to five strands of it. It is effective, economical, and durable. A barbed wire fence costs less than split rails, boards, or woven wire but probably more than electric fencing.

Woven Wire Fence

Woven wire makes a very gentle yet formidable fence. It is simply a net of wire with no sharp features. It can be used for all kinds of livestock and is particularly good for sheep because it has no barbs to catch the wool. It has a tight grid pattern close to the ground that makes it ideal for containing small animals such as hogs. Its height of four feet makes it quite adequate for cows and horses. Because woven wire has no electric shock or barbs to discourage animals, it may be subject to more stress than other types of fence. To discourage larger animals from leaning over the fence and bending it, string a strand of barbed wire along the top of the post above the woven wire.

As with barbed wire, there are variables to consider. In addition to choosing the gauge and the protective coating, you must choose the type of grid pattern you want. You need to understand how to decipher the code number. The last two numbers tell you the height of the fence in inches, and first two numbers tell you the total number of horizontal wires in the pattern. For example:

Style 1155. The fence is fifty-five inches tall and there are eleven wires.

Style 726. Stands twenty-six inches with seven wires.

Electric Fences

For many years barbed wire had no competition as the most economical and popular type of livestock fencing. Like barbed wire, electric fencing was another United States invention that found its way around the world. After being patented here, it was exported to New Zealand in 1937. Today, ironically, New Zealand and Australia are two of the major suppliers of electric fencing materials to the United States.

To be truly effective, livestock should learn to respect the electric wire. Feed should be placed around and under a live wire so that the animal will contact it once or twice and learn to avoid it in the future.

Controllers. The biggest cost in electric fencing is the controller, also called fence charger, fencer, or energizer. Its job is to send pulses of electricity into the wire. Some units allow you to choose how often you wish to send the pulse through the line. Usually once a second is sufficient for the livestock education period, then once every two seconds is enough. The intensity of the pulse varies with the type of unit. The very high-voltage, low-duration burst capacity of some units makes it almost impossible for the fence to be grounded out by high grass and weeds. Even though the voltage per burst is very high (five kilovolts) it is still safe because the duration of the burst is very short (2/10,000 of a second) compared to the other models (1/10 of second).

The two basic energy sources for controllers are batteries and household current. Units that require batteries are cheaper to purchase but are more expensive to maintain because the six-volt dry cell or wet cell must be changed every four to six months. If you forget to change it, you are left with a strictly psychological fence. Wet cells can be recharged; dry cells cannot. One new and exciting way of recharging wet cells is to attach a solar-powered photovoltaic unit on top of the controller. These have the potential of recharging the battery on sunny days. A fully charged battery can last three weeks even without any sunny days.

Wire. Usually a fourteen-gauge wire is used to carry the current. Heavier gauge (a lower gauge number) wires are more expensive but carry current with less resistance. Some imported systems use very tight spring-loaded wires which reduce the need for posts to about one every 150 feet. This wire is also usually fourteen gauge.

Almost all electric fence wires are galvanized. This is simply a zinc coating which makes the wire last longer. Some wires are more galvanized than others. A class III galvanization means the wire has a greater amount of zinc coating and therefore a greater durability.

Polywire and Netting. Polywire is a polypropylene cord with filaments of conductive metal wrapped in it. The colorful highly visible cord is easy to unroll and hang on movable posts. Polywire lines make excellent temporary fencing. A prefabricated net of polywire can also be hung from plastic posts which look like small javelins stuck in the ground. Sheep and hogs can be enclosed quite simply with these electrified nets.

The Ground. We often dismiss the "ground" wire of an appliance as just an additional bother. But, without a good grounding rod you have no electricity in the wire. The current must travel from the controller to the wire to the animal to the ground and back to the controller if the animal is to feel the shock. When an animal touches a live wire it essentially completes the electrical circuit. By allowing electricity to pass through it from the wire to the ground, it feels

Amount of Zinc Coating on Wire (ounces/square foot)

	14½ Gauge	9 Gauge
Class 1 (good) Galvanization	0.20	0.40
Class II (better) Galvanization	0.40	0.60
Class III (best) Galvanization	0.60	0.80

"shocked." However, a bird that sits on the wire without touching the ground does not complete a circuit and is unaffected.

Contrary to popular belief, the fence wire does not have to make a circle for the fence to work. You can run a wire out from the controller in a straight line for a mile and it will still do the job.

The best ground is a copper tube planted six feet deep in soggy soil. Copper is expensive, so you can settle for a copper-coated steel pipe or even any good conducting metal rod. Some electric fence manufacturers recommend the use of two or three of these grounding rods. They want you to take grounding seriously.

Construction costs for electric fencing are usually significantly cheaper than for any other type of fencing. Frequently only one strand of wire is necessary, especially if you are only supplementing an older fence. Wire stretchers and other additional construction equipment items are usually not needed.

The versatility and economy of the newer electric fencing materials make this type of fence a growing favorite of people who need a thrifty, practical, and hardworking fence.

Picket Fences

The all-American white picket fence carries all the neatness and formality of white gloves. Construction of a handsome picket fence requires patience, precision, and an attention to detail such as constantly measuring little distances and checking alignments with a bubble level. The designs of pre-cut pickets available in most lumber yards are quite limited. With a little extra effort and imagination, you can customize a picket top and have it cut out by a lumber yard. You could also do it yourself with a C clamp and a saber saw.

The cost of constructing a picket fence is roughly comparable to that of a split-rail fence. The picket fence is mainly a decorative item. Its low height and openness make it a friendly and attractive boundary statement that deflects wandering pets and people but not motivated intruders. The picket fence also seems to have a unique capacity for unlimited absorption of all of the creativity that you can give it.

How to Select the Right Tools

Some of the tools used in constructing fences are simple and yet ingenious. You should be familiar with the ones listed below so that you can make your fencing job as easy as possible.

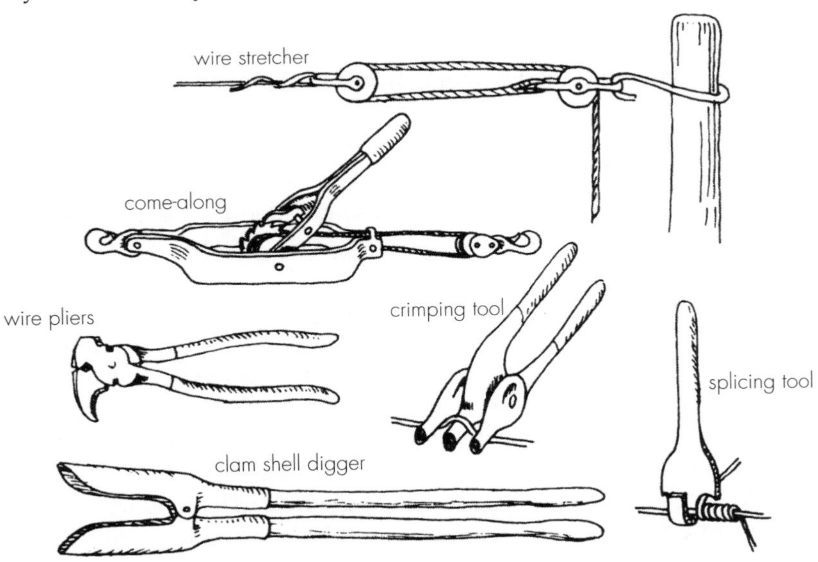

The metal post driver. A metal post is very hard to hit with a sledgehammer because the top is so small. The metal post driver is a heavy pipe welded on the sides so that you can pick it up and drop it a few times on the fence post. If you have trouble finding one to buy, you can have one welded for you locally.

Fence pliers/staple puller. This tool can do a million things. It can grasp wire, cut it, or twist it. It can dig out staples or hammer them in. It can even be used as a lever to stretch wire.

Wire stretcher/splicer. This is a device that has a greater leverage than pliers for the job of stretching a strand of wire. One end connects to the wire and the other attaches to a tree or dummy post for wire stretching or attaches to another wire for splicing.

Come-along. When stretching the multiple strands of wire in a woven wire fence, a comealong can be very useful. One or two of these can be attached to the dummy post or tree and then connected to the wire which has been clamped between two bolted boards. Two come-alongs make it much easier to distribute the tension between the top and the bottom of the fence, especially with bigger sizes of woven wire.

Wire splicer. This simple tool allows you to wind wire around itself easily. Some have holes with varying diameters to accommodate varying gauges of wire.

Nails. The most frequently used nails in fencing operations are called "common" nails and "box" nails. The common nails are sturdier. Both can be obtained with a galvanized coating which prevents rusting.

Carpenter's level. This metal device is at least one foot long. It has two bubble tubes set at 90 degrees from each other.

Plumb bob. This is a metal weight suspended on a string and is used to insure proper vertical alignment.

Wire Tighteners.

- *Crimping tool.* These funny-looking pliers deform wire in an attempt to tighten it. By making a straight wire wavy, they can take the sag out of the fences. The crimps, however, may not last very long and soon you may be back where you started. This device may be quite useful on older wire that can't take much bending or twisting.
- *Wire reels (Reel-tight).* The sagging wire can be wound around a pair of metal fingers until the desired tightness is reached. These sturdy metal prongs come in many sizes and are a very fast, effective method for eliminating sags, except in older wire that may break easily.

Staples. These sharp U-shaped fasteners come in various lengths and gauges. Contrary to popular practice wire should not be mashed into the post with staples, but rather held close to the posts allowing for some movement of the wire. For very hard wood posts, such as locust, use very short, thin staples.

Hinges. Four kinds of hinges are used for light, medium, heavy, and very heavy tasks involving gates.

- *Butt end hinges* are used primarily for light jobs. The leaves of this hinge attach to interfacing surfaces between the gate and the gate post.
- *The strap hinge* is designed for stronger work and attaches to the front surface of both the gate and the gate post.
- *The T hinge* is a hybrid of the first two.
- *The lag-bolt strap hinge system* is the most useful for very heavy tasks.

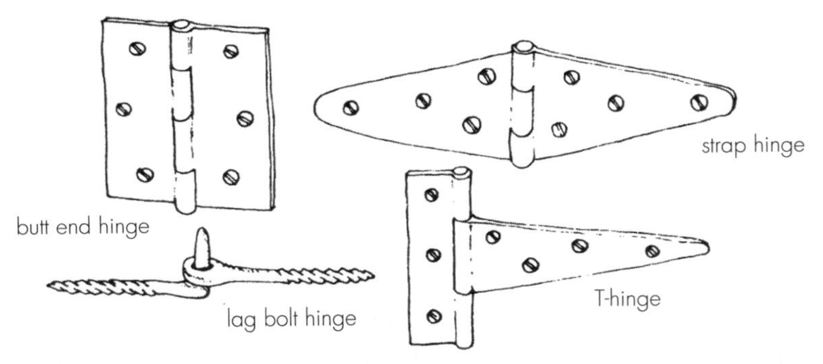

Latches and bolts. There are many different ways to close gates. The most common ones are thumb latches and dead bolts.

Hand auger. This devise allows one or two people to drill a beautifully cylindrical hole every time. The principle of this tool is the same as the corkscrew.

Power auger. This machine replaces people power with horse-power. Renting one of these hole-diggers for a day can save an enormous amount of time and backache, although the motorized nature of this tool does move you out of the frontiersman class temporarily.

Clam Shell Digger. This device is basically two narrow shovels hinged together. With the long handles held together, the clam digger is jammed into the soil. Sharp blades help a great deal here. Then the handles are pushed apart, trapping dirt between the blades. With a little twisting action, the unit is pulled out of the ground with its plug of dirt. The maneuver is repeated several times until you have created a nice, narrow, cylindrical hole about two feet deep. At this point, you can deepen the hole with your long iron bar, thrusting and stirring. The beauty of this technique is that it leaves the surrounding soil undisturbed, ready to be compressed by the driven post.

How to Use the Best Construction Technique

Good construction technique starts with stable, long-lasting fence posts. The steps involved vary with the type of fence you choose to build.

The Virginia Rail Fence

1. Lay out the fence with stakes and twine.

2. Set each bottom rail on solid flat rocks or concrete to keep it off the ground so it will not decay as quickly.

3. Lay adjacent rails at a 130-degree angle with an overlap of approximately one foot.

4. Continue stacking the rails one row at a time.

5. If the fence is not circular, the rails in the first and last sections can be set on the ground in a fan pattern, set in a mortised end post, or held between a pair of end posts supported on dowels.

The Split Rail Fence

Splitting the Timber. The first technique to master with this type of fence is splitting rails. It is usually possible to divide a 15-inch timber into four pieces quite easily. If you can then divide each of these poles in half again to get a total of eight poles, that's great. If you push to get sixteen poles you may end up with stockade fencing instead of split rails.

1. Some people advocate notching the butt end of the timber with a chainsaw along each of the diagonals that you intend to split. This step is not essential, but helpful.

2. Drive a steel wedge into the notch until splitting starts.

3. Drive additional wedges into the log until it splits. Continue splitting pieces to get four or eight rails.

4. Repeat the process with the remaining timbers.

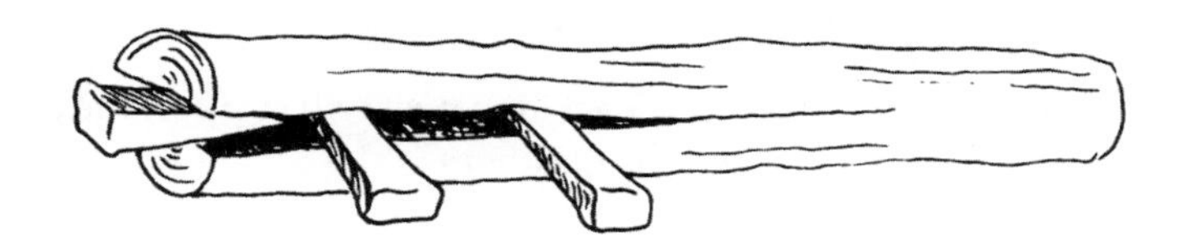

Mortising the Posts. Making slots through the posts to accept the rails is called mortising. It should be done before setting the posts in the ground. A two-rail fence, with two mortises per pole, is most common.

1. Mark the area of the wood to be removed. Measure down from the tops of the poles.

2. Use a two-inch drill to remove most of the wood, then complete the job with a mallet and chisel.

3. Taper the ends of the rails so they will fit together in the mortise. They can be overlapped either vertically or horizontally.

Setting out the Fence. 1. Plan on setting posts about two feet closer together than the length of the rails — six feet apart if you are using eight-foot rails.

2. Set one post firmly in place. Put the second post in the hold at the proper depth, so rails will be parallel to the ground. Put the two rails in position in both posts, get a tight fit of the rails, then tamp the second post into position. Repeat this process, each time putting rails in the second post before tamping it into position.

3. Heavy animals will rub against rails, loosen them, and eventually knock them down. To halt this, run a strand of barbed wire across the top rail.

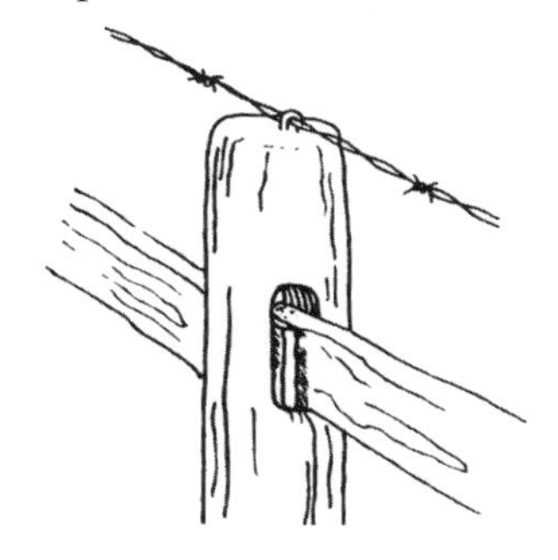

The Board Fence

1. Plan on setting posts a distance apart that equals half the length of the boards — six feet apart for twelve-foot boards, for example. In that way, each board is supported by three posts.

2. Arrange the boards so that each line post will support the center of some boards and the junctions of others.

3. Mark each post at the points where the top of each board will cross it.

4. As with the split-rail fence, place the first post firmly in position, then put the second in the hole at the proper depth, but don't tamp it into position until you've measured its exact position with the boards that will be nailed to it.

5. Place the boards on the same side of the posts as the livestock. This way the fence will be supported by the posts as well as the nails.

6. Nail the boards in place, being careful not to split the wood by nailing too close to the edge of the board. Also, stagger the pattern of nails so that they are not all in the same vertical line, to avoid splitting the posts.

7. Nail a small batten vertically over the board junctions at each line post for additional support and trim.

8. The top of each board should be beveled to shed water. A small block of wood can be used to cap each post, or an additional board can be run along the top of the beveled post. This creates an excellent "roof" for the posts.

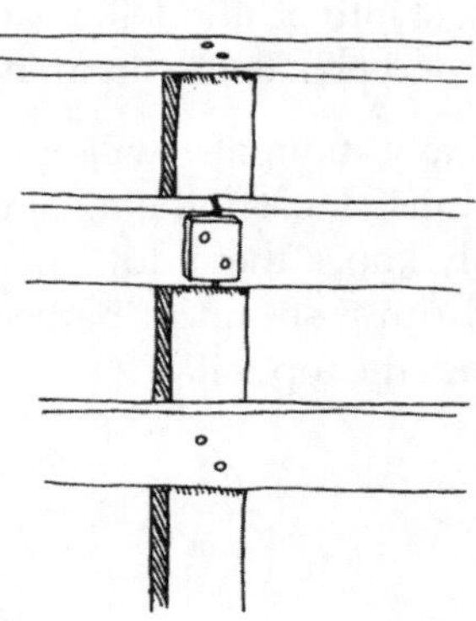

The Barbed Wire Fence

The principal job is to stretch the wire from the first to the last post in a straight line and then staple the wire on the posts in the middle. The posts in the middle don't contribute to the tightness of the wire; they keep it at the right height and provide stability in any situation that puts stress on the fence.

1. Set posts eight feet apart and brace the end posts. In some situations where stress on the fence will be slight, posts can be placed as much as sixteen feet apart.

2. Wind barbed wire around the first post and then wind it around itself, using the wire splicer.

3. Unroll the wire around the inside (your property side) of the line posts. This is easier if two people hold the ends of a crowbar which has been inserted through the hole of a bale of barbed wire, then walk down the fence line, letting the wire unreel.

4. Go beyond the last post to a dummy post or a tree which is in line with the line posts.

5. Attach the wire stretcher to the dummy post, then attach it to the barbed wire.

6. Stretch the wire cautiously by working the stretcher. Don't use a motor vehicle to stretch barbed wire.

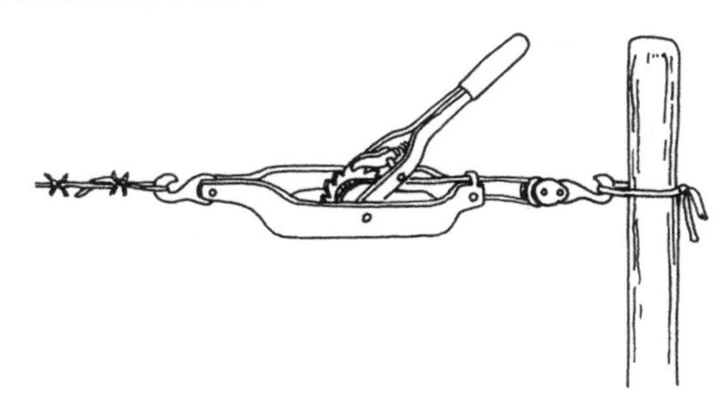

7. The wire has two strands. Cut one of them about two feet past the last post and remove any barbs.

8. Wind that strand around the post and then around itself. This allows the single wire strand on the stretcher to hold the tension of the whole line of the barbed wire while you are securing the other strand to the post.

9. Cut the remaining strand and repeat the wiring process.

10. Staple the wire to the line post. Avoid setting the staple legs in a straight vertical line because the post may split. Don't try to make the staple squish the wire into the post because this weakens the wire and ruins its protective coating.

The Woven Wire Fence

1. Set up a braced end post and line posts set eight feet apart.

2. Unroll the fence along the inside of the posts.

3. Make sure that the narrow grid portion is on the bottom of the fence posts.

4. Wrap each horizontal wire around the post and then around itself.

5. Do the top horizontal wire first, then the bottom horizontal wire, then the middle ones. You may have to cut a few vertical lines in order to get enough wire to work with. When the fence is fastened securely to the initial post, unroll the entire length of the fence.

6. Go beyond the braced end post to a dummy post or a tree and set up your wire stretchers. One way to do this is to bolt wooden slabs together through the fence and pull on the slabs with one or two or more wire stretchers or a come-along. Using the wooden slabs distributes the tension more evenly along the fencing and helps to avoid breaking the wire.

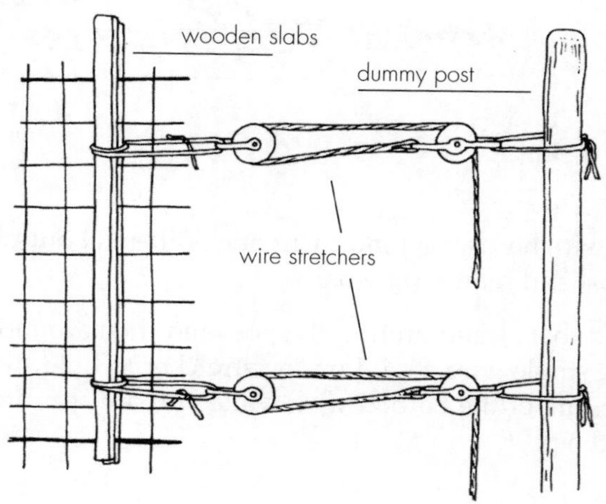

7. Stretch the fence carefully. Most woven fencing has built-in kinks along the horizontal wires. These help to maintain tightness. When you stretch the wire out, don't pull the wire so hard that you straighten out these kinks. Stop at the point where they just start to flatten a little. A few loosely placed staples holding the top wire to the line posts will hold the fence in the right position as you stretch. When the fencing is in place, start connecting it to the end anchor post, the one nearest the stretchers. Free enough of the central horizontal wire so that you can wrap it around the post and then around itself. Next, repeat the procedure with the horizontal wire halfway between the central wire and the top wire. The next wire to attach is the one that is halfway between the central wire and the bottom wire. Then do the remaining horizontal wires, saving the top one for last.

8. Go to each of the line posts and staple in the horizontal wires.

The Picket Fence

1. Plan the distance and materials on paper as discussed previously.

2. Place stakes in the ground at post sites six feet apart and then dig the holes.

3. Set the posts two feet deep on gravel and surround with concrete.

4. Check all alignments.

5. Wait two or three days before continuing the fence in order to allow the concrete to cure.

6. The wooden posts most commonly used measure four inches by four inches by five feet and have been treated with preservatives.

7. While you wait, cut the rails and the stringers and the pickets. A typical picket is one inch by three inches by three feet. The typical stringer is two inches by four inches. There are several ways to build the framework to hold the pickets. An easy method that avoids the dados and mortise joints calls for butt ending stringers along the tops of the posts with a mitered joint at each corner. The lower stringer should be supported by blocks and toenails.

8. Attach the pickets to the framework using a spacer bar that is a little narrower than the pickets. If the pickets are three inches wide, the spacer should be approximately two inches wide. Hang the cleated spacer bar from the top stringer to determine the exact place for each picket to be nailed. Use galvanized nails. When nailing the picket, it is helpful to also run a temporary board between the posts at the level of the bottom of the picket. This board keeps the bottom of the pickets on a level and will thus keep the tops aligned. Check the alignment frequently with the bubble level.

The Electric Fence

1. Establish sturdy corner posts that are braced.

2. Set a fence post every 150 feet if the terrain is level. If the land is uneven or if you can identify any area that must withstand unusual animal stress, use posts more frequently.

3. Unroll the wire from a flat reel set in the ground to work like a lazy susan. If you plan to have more than one strand of electrified fence, work with one strand at a time to avoid getting tangled up. The thinner the wire you choose, the tougher it is for your animals to see it. Make it noticeable by tying bright plastic ribbons on it, especially if it is sixteen gauge or thinner and you're only using one strand.

4. Attach the wire to the corner post with an end insulator.

5. Attach the wire to all the other insulators on the posts or on offset brackets, making sure the wire is allowed to slide freely at all points. If you are using one strand, set it thirty-six inches above ground. If you're using two strands, set one at seventeen inches and one at thirty-six inches.

6. Attach springs to the wire, if desired. Then by winding wire tighteners, the tension from the springs can be brought to about 200 pounds.

7. Use spring-loaded plastic handles at each gate area.

8. Plant the grounding rods properly. Then attach the controller to both the grounding rods and the fence wire.

How to Construct Workable Gates

Regardless of the style of fence you construct, you will need at least one gate. Figure on four feet for entering lawn equipment; up to sixteen feet for farm equipment. There are three areas that must be given careful attention in gate building to avoid the big mistakes.

- Foremost, the support post must be sturdy.
- The post itself should be bigger, taller, and set deeper than the remaining line posts. It is not uncommon to use a 6 x 6 or 8 x 8 preservative-treated post set at least three feet deep in a collar of concrete and braced well to an adjacent post.
- The gate must be hung properly. Sagging gates are ugly and irritating to use. If the gate is supported by blocks and guy wires while it is being hung, you won't have to fight the gate while attaching it to the rest of the fence. Flimsy hinges, improper bracing, and poor alignment can all be avoided easily.
- The gate must be structurally sound. A gate must be as light as possible and yet strong. This is the reason aluminum gates are so popular with farmers.
- A satisfactory wooden gate can be constructed with less cost by using good quality, dry, strong, pressure-treated wood, using correct bracing techniques, using the appropriate hardware, and by following the simple construction steps listed below.

Basic Gate Construction

1. Make sure you have very strong gate posts as mentioned previously for wide (twelve to sixteen feet) hanging gates. You may want to use a very tall gate post so that you can add support to the gate by running a cable from the top of the post to the latch side of the gate and tightening it with a turnbuckle.

2. Measure the opening for the gate. Measure it from top to bottom on both sides and measure the distance between the gate posts at the top and at the bottom to make sure that the gate is constructed as a true rectangle.

3. Working on a flat surface, construct a frame for the gate that is one inch shorter than the horizontal measurement made in the previous step so that the gate can swing without interruption. Using pressure-treated two-by-fours, make dado joints or rabbet joints for the corners. Also block these joints and use large wood screws.

4. Brace this basic frame with a strong diagonal strut. The alignment of this supporting structure is very important. The low end of this bracing bar must be placed on the hinged side of the gate, allowing the high end to be attached to the left side. For additional support, a cable and turnbuckle can also be used. To be effective, this system must be set on the opposite diagonal from the wooden brace.

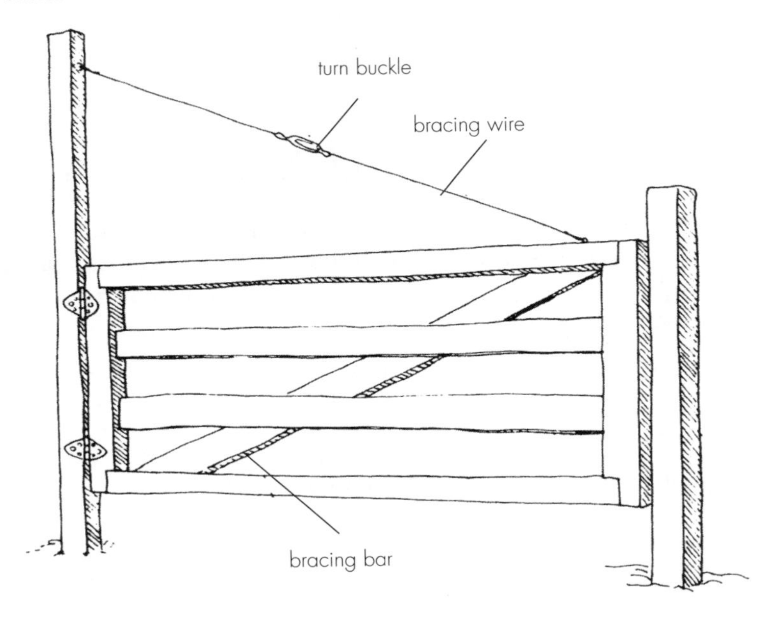

Following the placement of these bracing systems, the corners of the gate should be blocked with rectangular or triangular pieces of wood.

5. Nail the pickets or boards or other fencing trim in the proper pattern to match the remaining fence.

6. Set the gate on blocks in the same position you want it to be when you finish. Wire it into place temporarily with guy wires if necessary.

7. Attach the hinges by using large wood screws to attach the gate-side hinged leaf first. You should choose the largest wood screws that will go through the holes of the hinges but not through to the back side of the wood. Gates are usually too heavy for butt hinges, so strap hinges are usually used. For large gates, you will get much more long-lasting support from a lag and strap hinge system.

8. Attach the gate post hinge leaf with similar screws.

9. Undo the guy wires and remove the blocks. If necessary, trim the end of the gate so that it can swing freely for attaching the latch.

10. Set the latch or bolt system of your choice.

11. Place a stop-board on the post behind the latch.

How to Build Safely

Fencing materials and tools can be very dangerous. Barbs are sharp, electricity can kill, one strand of wire under tension can snap back with enough force to cripple you. Here is a list of some basic fencing dos and don'ts to remember.

Don'ts

- Don't carry staples in your mouth.
- Don't steady a post with your hands while someone else hits the top of it with a sledgehammer.
- Don't wear loose clothing around power augers and barbed wire.
- Don't let your feet or legs stay in the arc of the swing when you are pounding posts into the ground — just in case you miss.
- Don't shortcut a good controller in your electric fencing.
- Don't work with fencing materials in a lightning storm.
- Don't stretch wire with a tractor. If it breaks you are in big trouble.
- Don't try to clear the weeds and brush beneath a wire fence by burning it. The fire may weaken the posts and ruin the protective zinc coating on the wire.

Dos

- Wear some eye protection, especially when hammering staples into very hard wood, hitting a metal post with a metal maul, or breaking up rock in a post hole.
- Anticipate the whipping action of wire, especially when stretching a new line of wire or when unstapling an old line.
- Stay on the side of the post away from the wire.
- Avoid getting chemicals from treated fence posts onto your skin or into your eyes.
- Place a grounding rod somewhere along non-electric fencing unless you are using metal fence posts.
- Always wear gloves.

Suppliers

Kencove Fence Supplies
Blairsville, Pennsylvania
800-536-2683
www.kencove.com

Kiwi Fence
Blairsville, Pennsylvania
800-536-2683
www.kiwifence.com

Parker-McCrory Mfg. Co.
Kansas City, Missouri
816-221-2000
www.parmakusa.com

Premier 1 Supplies
Washington, Iowa
800-282-6631
www.premier1supplies.com

Southwest Power Fence and Livestock Equipment
San Antonio, Texas
800-221-0178
www.swpowerfence.com

Twin Mountain Fence
San Angelo, Texas
800-527-0990
www.twinmountainfence.com

Other Storey Titles You Will Enjoy

The Backyard Homestead Guide to Raising Farm Aniamals, edited by Gail Damerow.
Expert advice on raising healthy, happy, productive farm animals.
360 pages. Paper. ISBN 978-1-60342-969-6.

The Fence Bible, by Jeff Beneke.
A complete resource to build fences that enhance the landscape while fulfilling basic functions.
272 pages. Paper. ISBN 978-1-58017-530-2.

Fences for Pasture & Garden, by Gail Damerow.
Sound, up-to-date advice and instruction to make building fences a task anyone can tackle with confidence.
160 pages. Paper. ISBN 978-0-88266-753-9.

How to Build Animal Housing, by Carol Ekarius.
An all-inclusive guide to building shelters that meet animals' individual needs: barns, windbreaks, and shade structures, plus watering systems, feeders, chutes, stanchions, and more.
272 pages. Paper. ISBN 978-1-58017-527-2.

Making Bentwood Trellises, Arbors, Gates & Fences, by Jim Long.
A thorough guide to collecting limbs from native trees, then using them to craft a variety of designs for the garden.
160 pages. Paper. ISBN 978-1-58017-051-2.

Renovating Barns, Sheds & Outbuildings, by Nick Engler.
Step-by-step advice on how to square and strengthen the structure, repair or replace the roofing and siding, install new windows and doors, and even add electricity and plumbing.
256 pages. Paper. ISBN 978-1-58017-216-5.